SPACE DOCTORS

NAT THE NEUROSURGEON

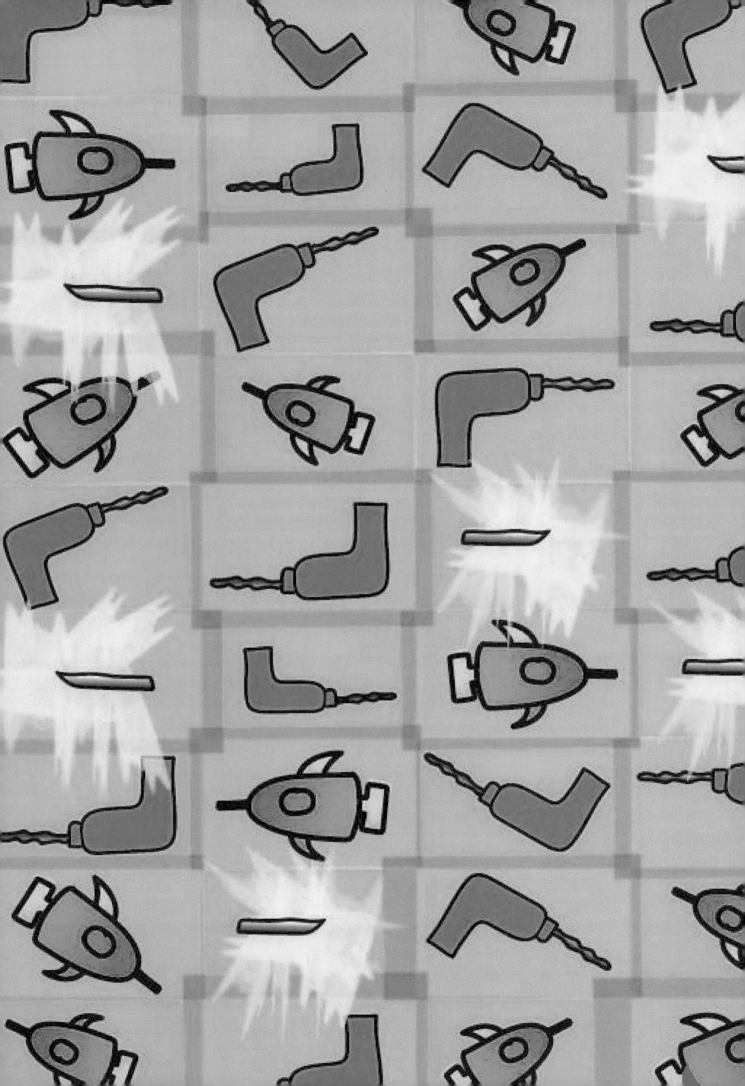

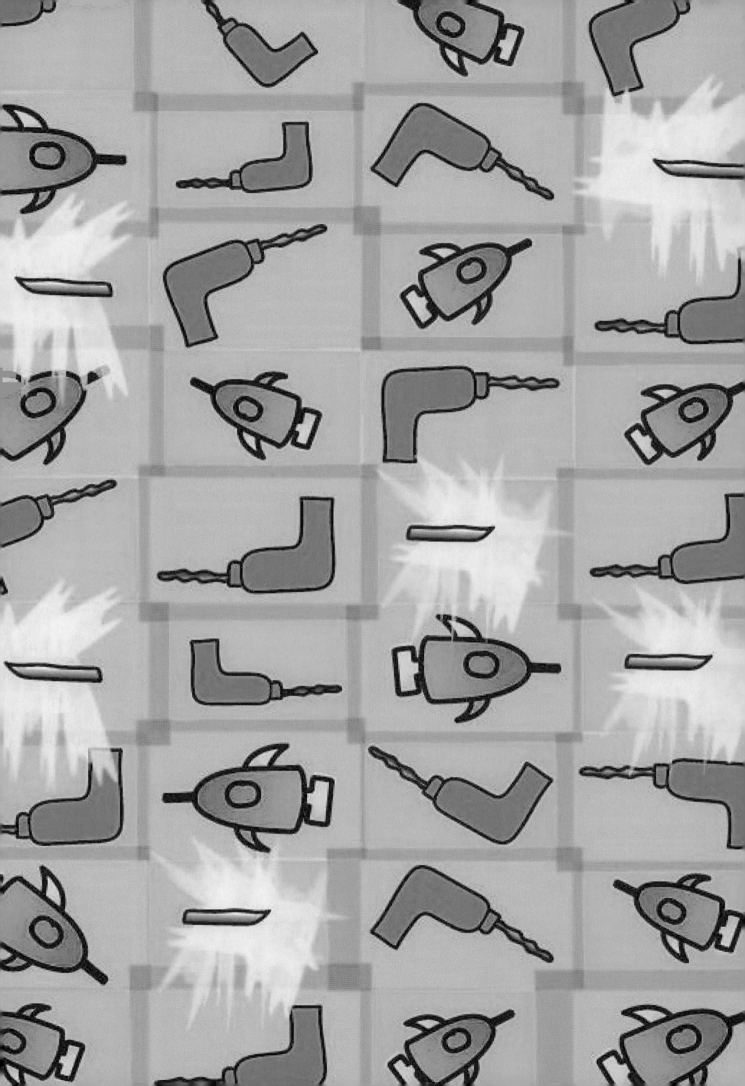

For F.S, L.A and A.T. - whose endless love of complicated vocabulary started us on this journey.

Meet
Neurosurgeon Nat,

And her trusty rocket Rover

She's a very special doctor from a distant supernova

Travelling the galaxy, alleviating pain

Of aliens and spacemen with problems in the brain.

She has lots of clever gadgets and a special surgeon's knife

And is always there on hand to save somebody's life.

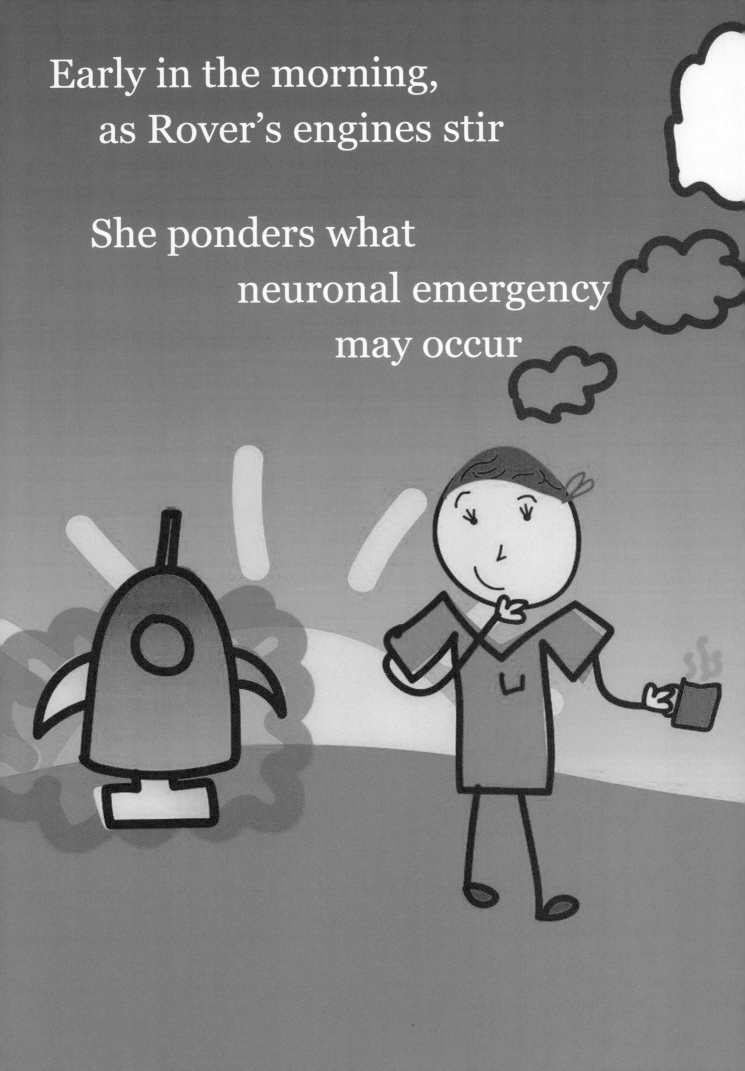

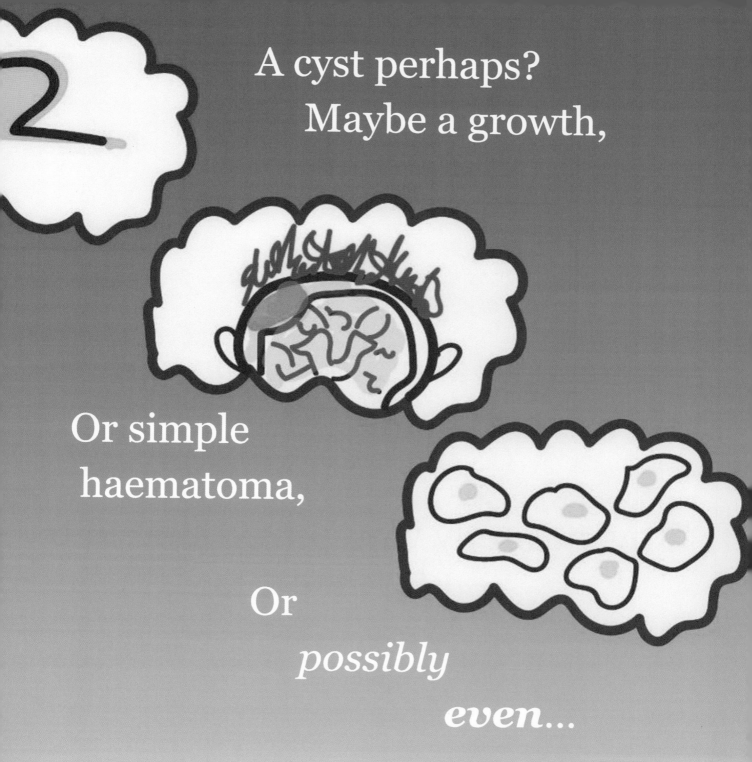

Zipping past a waning moon, she hears
an awful cry,

Rover nimbly whipping round to try
and find out why,

Touching down on dusty rock, she
glances round to see

A stumbling moon man, head in hands
and moaning,

Please help me!

'What seems to be the problem?'

Nat unpacks her surgeon's case

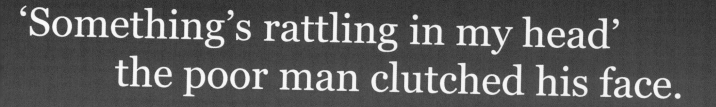

'Something's rattling in my head'
the poor man clutched his face.

She grabbed her scanner
and took a look,

the cause she learned
with ease,

'Your problem, she proclaimed, is that

your head is full of cheese!'

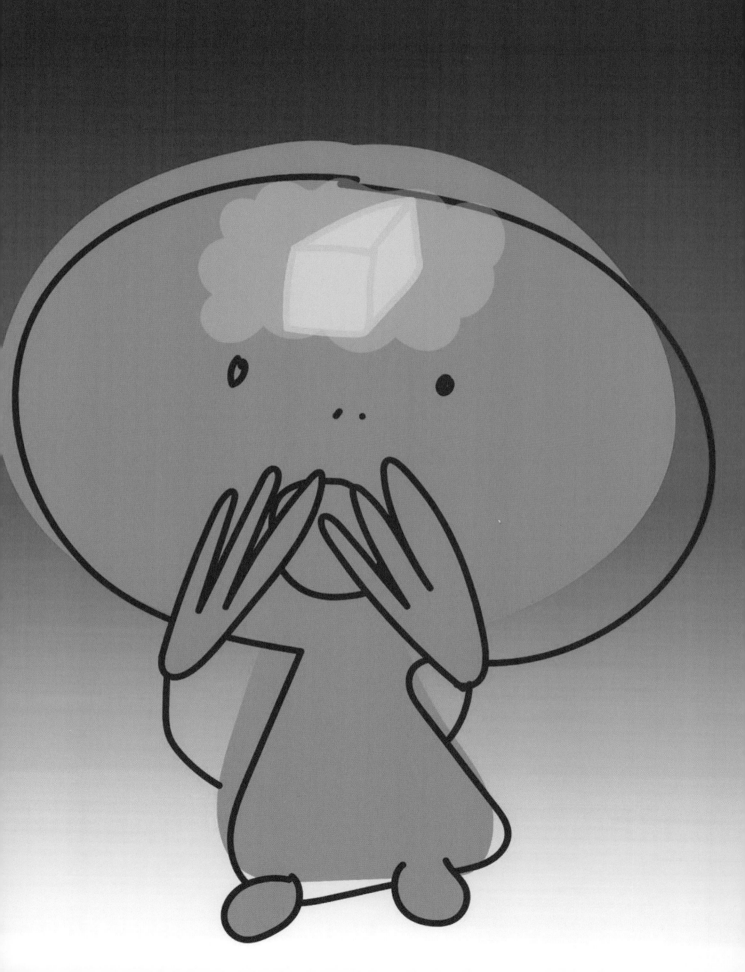

Natalie knew what to do, firing up her drill,

Into the skull and dura, with dexterity and skill.

With a flicker of delight,
'**Got it!**'
Natalie said,

And plucked the cheesy culprit from the
grateful moon man's head.

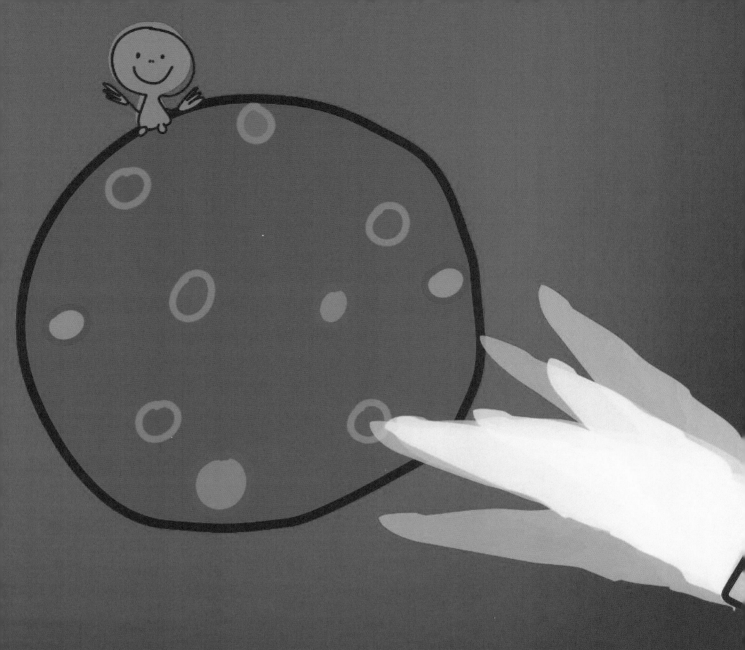

He thanked the surgeon graciously and skipped off with delight,

As Nat the neurosurgeon blasted off into the night.

"Another patient saved, eh Rover?"
Natalie exclaimed

"I think I may just have this little
piece of moon cheese framed!"

Zooming by a scarlet Mars, Nat had to look twice,

She thought she saw a massive head, but could not be precise.

Soaring in to take a peek, confirming what she knew

For once Nat wouldn't need a CT scan to give a clue.

Baffled martian doctors cried, "He continues to vex us!"

Nat replied "I think the problem's near his choroid plexus"

"Too much water on the brain, we call it hydrocephalus,

He needs a ventriculostomy, that alone will save us."

She grabbed her case and abseiled down from Rover on a rope,

And fired up her shiny neurosurgeon's endoscope.

She injected anaesthetic to take away his pain

Then advanced the scope smoothly pas the thalamostriate vein.

At the floor of his third ventricle, the
scope came to a stop

She made a tiny hole, and the fluid
drained with a pop

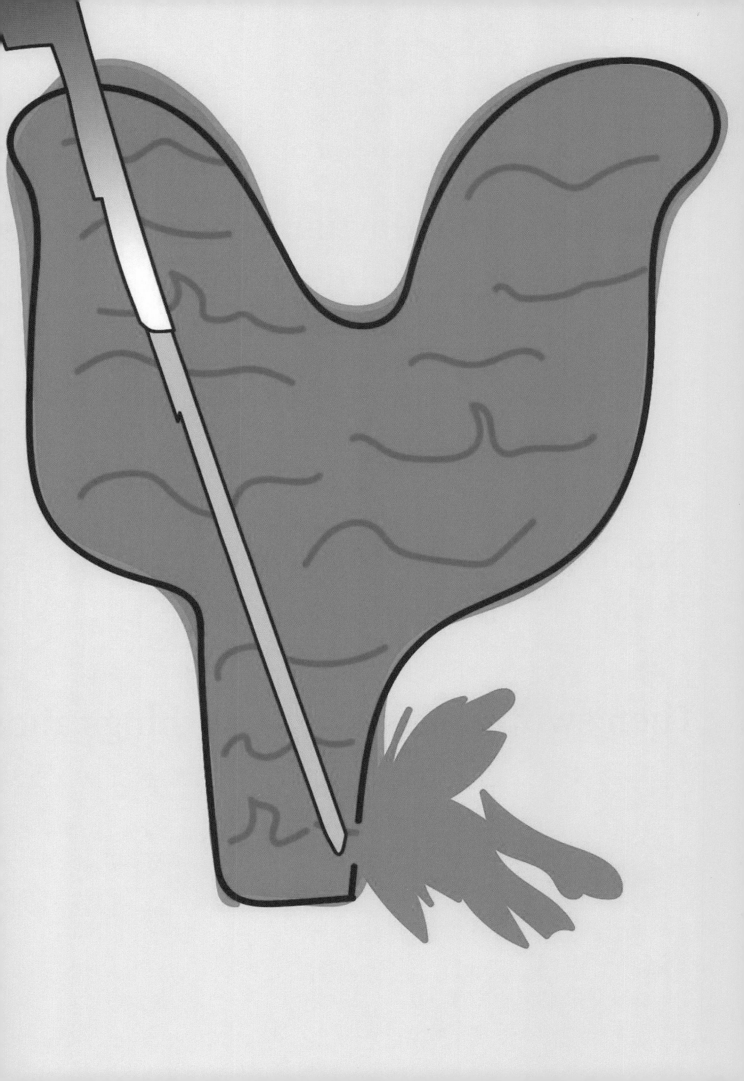

"**Hooray!**" the Martians cheered as Na
removed the scope

"It's brave Surgeons like Natalie that
give us Martians hope"

Natalie gave the Martian careful post
operative advice

Then swiftly packed up her things and
was off in a trice.

"Uh Oh, watch out!" blooped Rover, his dashboard gave a spark,

"My gosh what's that?" cried Natalie, peering out into the dark.

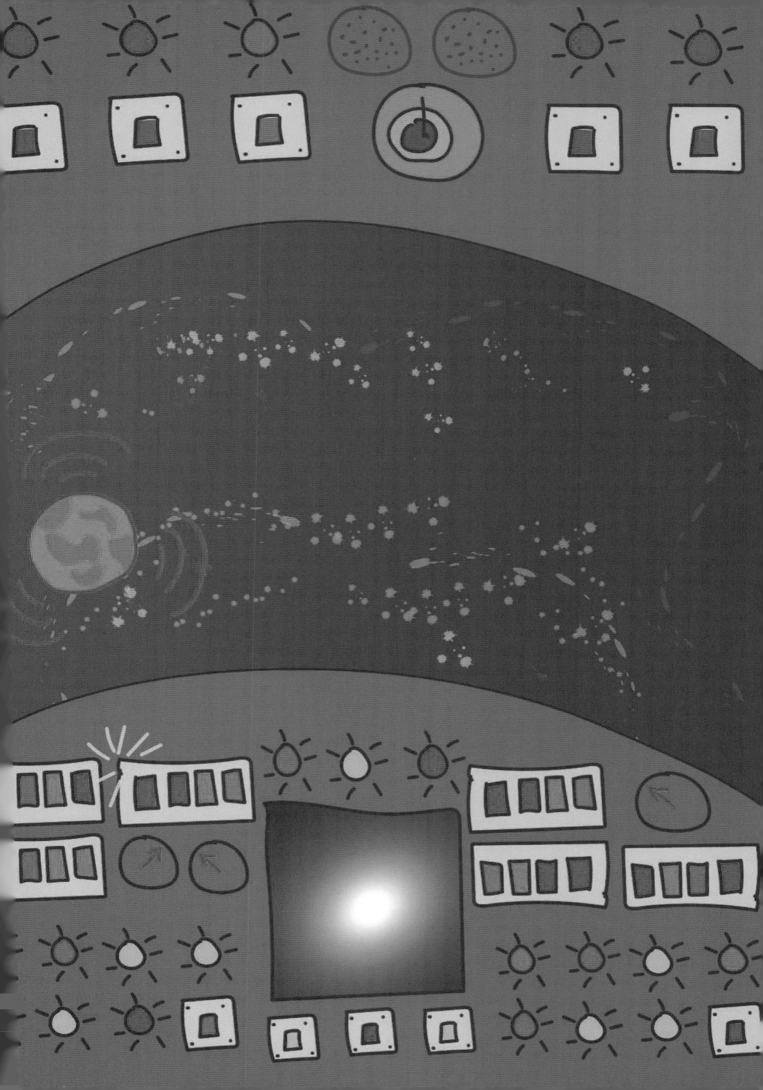

"It looks like
    one of Neptune's folk
    has slipped...

...and
bashed his head"

But he didn't
    yelp or cry...

...he simply lay
quite still instead

Baffled as to what to do, Nat proclaimed "Dear me,"

"I need to ask my friends for help at the neuro-MDT!"

To Ventriculon-5 the intergalactic
neurosurgeon flew

Hoping that her friends will help her
work out what to do.

There's Andy the anaesthetist,

Roland from radiology,

And Nitin with his microscope from team neuropathology.

They looked at blood tests, films and
scans
And conjured up some clever plans

Roland and Nitin argued
but finally agreed...

"This alien has had an intracerebral
bleed."

Nat scrubs in and dons her tools, and
once inside the brain,

She schlorps away the haematoma and
then inserts a drain.

"Fantastic work, a great success" cried Andy with a smile,

And to the Neptunian said, "you'll feel woozy for a while"

As the day draws to a close
Nat contemplates some sleep.

But as her head lands on the bed she's jolted by her bleep.

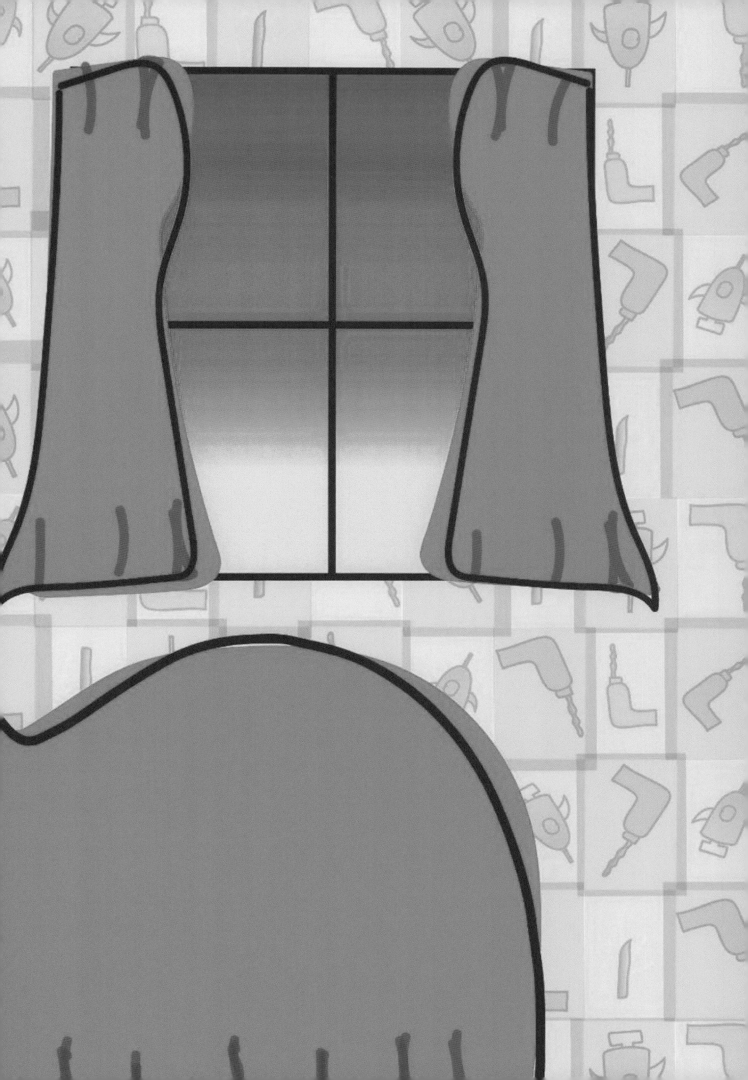

She sighs and smiles and rising up she takes the call with glee

Because she knows she'd **never** choose another specialty.

Glossary

Cyst - *A closed sac of fluid that can form anywhere in the body*

Haematoma - *A collection of blood inside the body that occurs due to injury to blood vessels*

Oligodendroglioma - *a type of brain tumour that develops from oligodendrocytes. These are a type of glial cell found in the brain. Glial cells are typically viewed as the cells which support the neurons of the brain (although some argue for a much more crucial role that simply 'supporting').*

Dura - *a protective membrane that surrounds the brain. It is one of three layers collectively known as the meninges.*

CT scan - *a CT or computed tomography scan uses x-rays to make a three-dimensional representation of the inside of the body.*

Choroid plexus - *an area of the brain inside the ventricles (see later) that is responsible for producing the fluid inside our brains know as cerebral spinal fluid.*

Glossary [continued]

Hydrocephalus - *a condition in which there is too much fluid within the brain*

Ventriculostomy - *a procedure in which you make a hole in the wall of the ventricles (see later)*

Endoscope - *a piece of surgical equipment which can be used to see inside the brain and perform operations with very small instruments. It has a camera, a light and a channel to pass instruments down. It causes less damage to surrounding tissues but is more technically challenging to operate with in the deep areas of the brain due to a limited range of movement.*

Thalamostriate vein - *a larger vein (blood vessel which carries blood with less oxygen back to the heart) found near two important structures - the thalamus and the striatum. The thalamus is like a relay station for sensory information (vision, hearing, smell etc) and the striatum is involved in coordinated movement.*

Glossary [continued]

Ventricles - *the ventricles are chambers within the brain which contain fluid (cerebral spinal fluid or CSF). There are four main chambers: the two lateral ventricles, the third and the fourth ventricle.*

Neuro-MDT - *MDT stands for multi-disciplinary team. The neuro-MDT is a group of medical professionals who have different roles (for example specialise in imaging, medical treatment, surgery, cancer). They come together to decide the best way to treat individual patients.*

Anaesthetist - *a doctor who puts people to sleep for operations and manages different types of pain. Check out the next book in our series, 'Andy the Anaesthetist' to find out more about what an anaesthetist does.*

Radiology - *branch of medicine that uses imaging technology to diagnose and treat disease*

Glossary [continued]

Neuropathology - *Pathology is a branch of medicine that focuses on understand cause of disease, largely by examining samples in a laboratory. Neuropathologists are concerned with diseases of the brain.*

Intracerebral bleed - *a bleed inside the brain that occurs when there is damage to a blood vessel.*

Scrubs - *uniform worn mainly by healthcare workers in operating theatres.*

Bleep - *an old fashioned device that doctors use to communicate.*

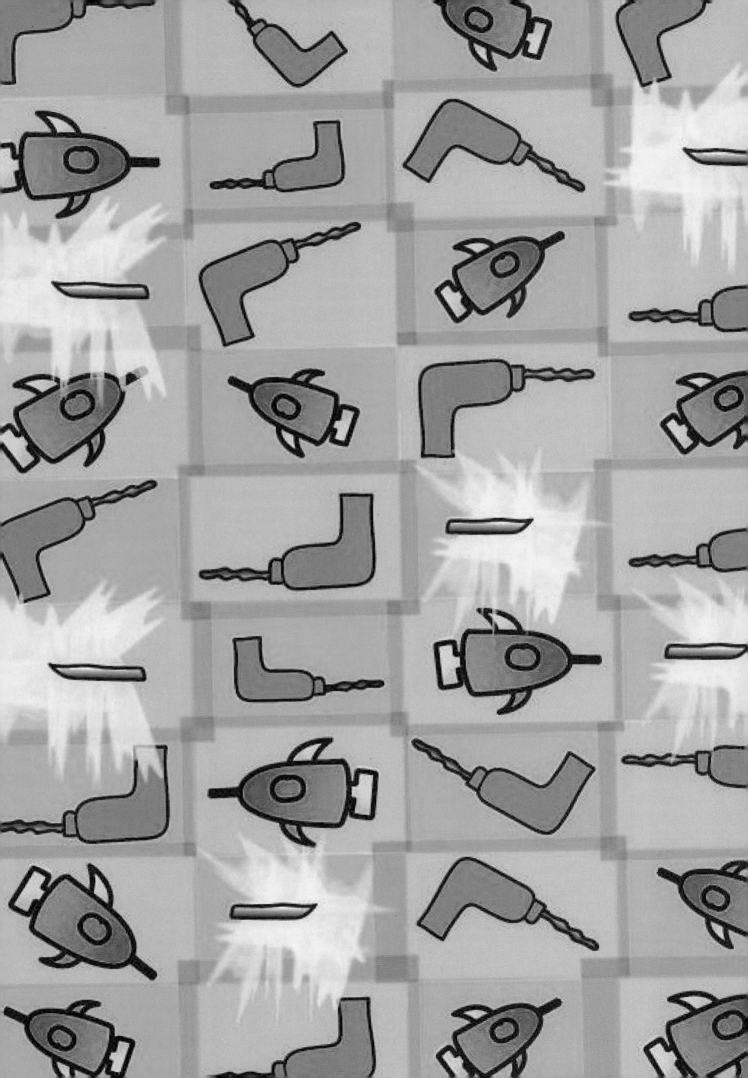

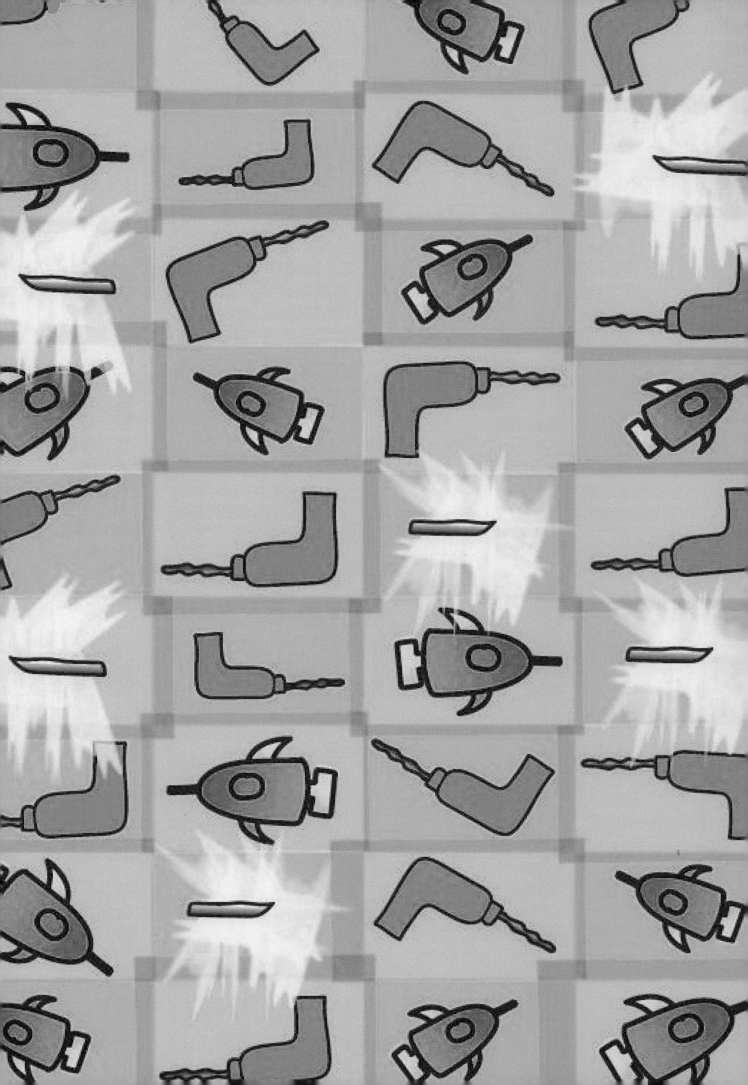

Printed in Great Britain
by Amazon